baby BOBCATS

KIM THOMPSON

CREATIVE EDUCATION • CREATIVE PAPERBACKS

CONT

I Am a Kitten 4

In a Den 6

Hunting at Night 8

Growing Up 10

Speak and Listen 12

Bobcat Words 14

Reading Corner 15

Index 16

I AM A KITTEN.

I am a baby bobcat.

I am named for my short, "bobbed" tail.

I was born in a <u>den</u>. My mom had three kittens at once.

I drink my mom's milk. After one month, she brings fresh prey for me to eat.

My mom teaches me to hunt rodents and birds. I jump high. I pounce! I kill with one bite.

I am nocturnal. I can see in the dark.

I am one year old. I am ready to leave my mom. I will live alone.

I will grow twice as big as a housecat.

SPEAK AND LISTEN

YE

OWWW!

Can you speak like a kitten?

Baby bobcats mew and chirp.

Listen to these sounds:

https://www.youtube.com/watch?v=JHP60sKuClc

Now it is your turn!

BOBCAT WORDS

den: a bobcat home in a hollow tree or cave

nocturnal: awake all night

prey: animals that are killed and eaten by other animals

rodents: small mammals with gnawing teeth such as mice and squirrels